WILD WISDOM

GERMAINE SMITH

THE CENTER FOR HEALING PRESS

© 2024 by Germaine Smith. All rights reserved.
ISBN : 978-0-9960421-6-1 (print).
Library of Congress Control Number: 2024907270

All rights reserved. No portion of this publication may be reproduced or transmitted in any form or by any means, electronic or mechanical, including photocopying, recording, or capturing on any information storage and retrieval system, without permission in writing from the publisher, except by a reviewer who may quote brief passages in a critical article or review to be printed in a magazine, newspaper, or electronically transmitted on radio, television, or the Internet.

The Center For Healing Press is the imprint of The Center For Healing, a spirituality center of education for those seeking wholeness.
For reprint permission: germ@thecenterforhealing.us.

Front Cover picture of Ludington, Michigan by Lois Tschida

All rights reserved.

Dedicated to
Vienna Giles Blacksher,
whose presence
will help the world

and

Mary McPherson,
whose presence
always helps me.

Overwhelming Thanks
to
Connor Blacksher
Mary McPherson
Lois Tschida

*Everything in creation
is alive
with Wisdom*

and

*Wisdom speaks through
everything
and everyone.*

Listen!

*Listen
to the
Wisdom*

everywhere.

Welcome to the World, Adorable Acorn!

You are a gorgeously unique,
    mysterious wonder of Creation.
Just like me.

I am part of the lemur family
    and
    am the largest nocturnal primate.
In fact, when humans first saw me
    in my native Madagascar,
    they were struck
    by my distinctiveness.

So it will be with you.
The world awaits you and your gifts!

Affectionately,
Allyn, the Aye-aye

Aho!

I may be tiny but I am very strong.
Just like you!

I can lift fifty times my body weight.
Thus, I can carry objects
      a very, very long distance.
And when my friends help me —
      Wow! We are a force.

My lesson is: you are strong.
Use it.
Trust it.
And when needed,
      call your friends.

Admiringly,
Aaron, Alan, Andrea, and Anthony,
      the Ants

Awesome Acolyte!

I am an amphibian called an axolotl
 who lives around Mexico City.

Although I am almost extinct,
 I am very important because,
 if damaged,
 I can regenerate,
  not only my limbs,
  but my heart, eyes, and brain.

So can you.
Perhaps not as much as I,
 but you too can heal yourself
 when you get hurt.

Embrace your ability to heal.

Accepting you always,
Adalyn, Agnes, Alice, Alexis,
Annika, Amanda, Amy, and Ashley,
 the Axolotls

B

Bonjour Beautiful!

I am an image of
    cuddly, ferocious protectiveness.

I am fierce.
I know when to stand on my back legs
    and roar.

I am content.
I know when to lay on my back
    and eat strawberries.

Trust your intuitive timing.
You know when to stand up
    to help someone.
You know when to lay back
    and enjoy life.

Baking in the beauty of becoming,
Brenda, the Brown Bear

Beaming Beauty!

I look like a rabbit but a little different.
My very long ears help me hear.

I hope you will listen to my plea.
Please treat everyone
     and everything
         with care and compassion.

I had cousins but they are now extinct.
I am the only one left in Australia.

So, I ask you to respect the land
     and all her creatures.
Then together,
     we can all live a long time
     on our beautiful planet.

Bursting with a bountiful heart,
Ben, Bob, and Brian,
     the Bilbies

Bedazzling Babe!

I am a member of the dolphin family
	called a Baiji.

There are very few of us
	in the Yangtze River pf China
	because we are almost extinct.
Getting caught in fishing nets,
	we cannot surface to get air.
Hence, we drown.

Remember, your actions reach far.
Everything is connected to everything!
What you do, affects everyone else.

Blessings,
Barb, Beth, Bethany, Blanca,
Brita, Britt, and Brynn,
	the Baijis

Charming Child,

I bring you gift of passion.

You have a zest for adventure
    that will help you
      reach your spiritual depth.

My bright red color reflects
    positivity and passion.
May your bright personality
    reflect your optimism
      as you explore your passions.

Captivatingly carefree,
Callie, Cathy, Christine, and Colleen,
    the Cardinals

Curious Cookie!

I am the traveling animal!
I journey over great distances
    without needing to stop
    for water or rest,
    easily carrying all necessities
    with me on my back.

I have found the world a beautiful place.
And so will you.

Be adventuresome.
Seek new places and peoples.
Travel!

Chao,
Connor and Corey,
    the Camels

Cheerful Cherub!

We share the same gifts.

Both of us are
    wise and
    independent and
    self- sufficient and
    unflappable and
    persistent and
    curious.

We can't be stopped!

Cordially,
Cal, Christopher, and Chris,
    the Cats

D

Dear Destiny!

My gift to you is the reminder
    you have the power
        to accomplish
        all within your sights.

You have the tools of ingenuity.
You have the knowledge of wisdom.
You have the connection of community.

While being a friend to all,
    you never lose
    your sense of self.

You go, Girl!!

Dutifully,
Denise, the Great Dane

Darling Daffodil!

I am a small antelope who lives in Africa.

I share with you the gifts of
    awareness and perception.
We are both creatures with pure hearts
    that lead with gentleness
    in a sometimes, harsh world.

Trust your insight and intuition!

Dreaming of you,
Dennis, the Dik-dik

Delightful Doll!

I am a fish who just loves the water.
I bet you do too.

I come in many colors,
    am very low-maintenance,
    and have eye-catching energy.
I bet your parents say
    the same things about you!

Darting, diving, and daringly yours,
Doris and Dina, the Danios

E

Exceptional Eagerness!

I am an Eagle
    known as
    "Chief of all flying creatures"
    because I fly the highest.

I encourage you to spread your wings.
Soar.

I am a symbol of the Great Spirit
    that lives in all creation.
As you soar to the heights
    know
    we are One.

Eloquently,
Emily and Emerald, the Eagles

Extra! Extra! You are Extraordinary!

I am called "electric"
    for a very good reason:
I can emit a shock of 800 volts!
My charge can burn your skin
    and knock you out.

So, my message to you is:
    Shock the world!
Be the extraordinary being you are.
We can be electric together!

Eternally,
Ellie, Eli, Ero, and Evan,
    the Eels

Energetic Elegance!

I am a grand animal in this world.
I symbolize compassionate leadership.

Draw from the Wisdom within you
— the Sacred Feminine —
to guide yourself
and your people
with
great understanding
and in harmony.

Earnestly,
Elias. the Elephant

F

Fancy Favorite!

As a Fangletooth Fish,
    I not very attractive.
I have a condensed body,
    massive head,
    protruding jaws,
    and grotesque appearance.

I live at the great depth bathyal zone,
    or "the midnight zone"
    of all the oceans.

Follow me.
Go deep.
Fearlessly enter where few venture.
You will be amazed at what you find.

Fervently faithful,
Frances and Francis,
    the Fangletooths

Fantastic Friend!

Life is hard, gentle one.
That is a fact.
I implore you to find ways
    to be seen and herd
    despite the risks.

Be open.
Be vulnerable.
 Be real.

I am called a glass frog
    because
        I have no coloration on my belly.
So, everyone can see right through me.
It's a bit scary but
    authenticity is worth it!!

Forever,
Fernando, the Glass Frog

Formidable Fire!

As a bird of prey, I am
    the fastest animal in the world.
When I dive for my meal,
    I reach 200 miles per hour.
In addition to my speed,
    I am incredibly precise.

I offer you clarity of mind,
    the need for attentiveness,
    and the ability to prepare.

Then you can overcome anything.

Flamboyantly fixated on you,
Florence, the Falcon

G

Glorious Golden Girl!!

I am a giant in the animal world!
Being the tallest of all land beings
    I stand 14 to 18 feet.
My neck alone is 6 feet long.

Despite my size, I am very graceful.
And that is my message to you.

Stand tall, no matter what your size.
Stick your neck out when needed.
And be graceful in all you do.

Gratefully,
Gratitude Group, the Giraffes

Glad to meet you, Gorgeous!

I am the largest of all primates and
    despite how the movies depict me,
    I am gentle and shy.

Sharing 98 % of my DNA with humans
    I am your closest animal relative.

And yes, I am very smart.
Not only can I use tools
    I can learn sign language
    to communicate.

I bring that gift to you.
Communicate clearly
    all that you think and feel.
Embrace your vulnerability.
Trust that there is always
    a nonviolent, peaceful solution.

Gleefully,
Greg, the Gorilla

Greetings Great Gal!

My lineage extents 157 million years ago
    to the late Jurassic Period.
I am a Gar, a freshwater fish
    with a skinny body,
    long rows of very sharp teeth
    and solid diamond scales.

Throughout history,
    my scales have been used for
    jewelry, hammers, plows, arrows,
    and armor.

So, be like me.
Be ready for anything.
You never know what purpose
    you will serve!

Guaranteed to help you anytime,
Germ, the Gar

Hello Hero!

I am one of the smallest birds
        and one of the fastest.

Learn the elegance of navigating life.

One way I do this is by going in reverse!
It's an important skill —
        to know when to go forward
        and when to go backward.

There is nothing wrong
        with going in reverse.
If fact, sometimes
        it is the smartest thing to do!

With that knowledge and skill,
        we can do anything!

Honorably,
Holly, the Hummingbird

Hooray Hopeful Heart!

I am here to teach you
      that how someone looks
      can be deceiving.

I am extremely large.
I can weigh up to 4,000 pounds.
And I look slow and very awkward.

But I can outrun any human!
I have been clocked at 30 miles an hour.

So don't judge a hippo by his looks.
Get to know us first!

Huge hugs,
Harold, the Hippopotamus

Howdy Honey!

I do have a strange name: Hallucigenia.
I am one of the earliest life forms
      in the earth's prehistoric ocean.
My ancestors lived 500 million years ago.
Today I call home the Arctic Ocean.

My Wisdom to you is: keep moving.
I have ten pairs of legs.
Yes, ten!
They propel me forward
      throughout the Arctic waters.

Wherever you are, whatever you do,
      keep moving.
This means facing all obstacles.
No matter what!

Humbly,
Hannah, the Hallucigenia

I

Impressive Imp

I am an Indris,
      a lemur who lives in Madagascar.

I am very cute
      with my big eyes, soft fur,
      and a gentle spirit.
I symbolize playfulness and positivity.

Remember to laugh and play every day.
Don't take life too seriously.

Finding joy in relationship,
      we develop warmth, connection,
      and lively camaraderie.

Whatever the crisis,
      laughter and companionship
      will help you through.

Industriously,
Iris, the Indris

Important Idol!

I am a member of the caterpillar family
    called an Inchworm.

I get my name because of my movement.
Unlike everyone else in my family,
    I don't have legs
      in the middle of my body.
So, I have to "inch" along.

That is my paradoxical Wisdom to you.
Take as much time as you need
    and
      move as fast as you can.[1]

Invigorated by your Presence,
Issac, the Inchworm

---

[1] Mary Palmer.

Imaginative Intelligence,

As an Impala, I am related to
    cattle, buffalo, sheep, and goats.

I am slender, lengthy, and very graceful.
A strange fact:
    I have scent glands in my ankles.

My greatest gift is my ability to leap.
I can jump 30 feet in length
    and 10 feet high!

There will be many times in life
    you will have to leap.
Sometimes in joy.
Sometimes to overcome hurdles.
Regardless of the reason,
    be graceful.
Trust your legs and the rest of you!

Inspired,
Isabel, the Impala

J

Joyous Joker!

I am the largest cat
    and the only "big cat"
    found on the American continent.

Often, I have yellow fur with black spots
    but can also be white or black.
Even then, I have spots
    though they are very faint.

But I am not your average kitty cat!

Originating over 3 million years ago
    I have survived in part
    due to my title of
    "strongest bite" of all cats.

My Wisdom: hang on with all your might.
Life will always get better
    if you persevere.

Jamming with joy,
Jana, the Jaguar

Jeweled Jade,

Jerboa is my name.
I am a small rodent
    from the deserts of Africa.
But I don't look much like a mouse.

I have long floppy ears like a rabbit
    and weird adaptive legs
      that allow me to
        jump like a kangaroo.

So, here's the truth!
It doesn't matter what you look like
    or what your name is.

Be yourself!

Jazzed to play with you,
Jagdeo, Jake, Jason, Jed, Jim, Joe,
JoJo, Joseph, and Jordan,
    the Jerboas

Jolly Jot!

I am a bird
    of Central and South America
    called a Jubiru.

I am a stork with
    very long pouch-like beak
    red throat patch
    huge white body
    and stick-like long legs.

Relationships are important to me
    as I mate for life.

My Wisdom to you is:
    friends, partners, family are vital.
Choose wisely.

Jocularly,
Janet, Jeannie, Jen, Jenna,
Julie, and Judy,
    the Jubiru

K

Killer Knockout!

Sometimes, one must realize
>how power can inflict pain.

I am one of the three "Big Snakes"
>along with the Cobra and the Viper.

When I bite, the bite is often painless.
But usually, deadly.
My distinctive yellow and black bands
>do serve as a warning.

Here's my Wisdom.
Be aware of your power to inflict pain.
Then give lots of warnings
>before you bite.

Keenly,
Kate, Kendra, Kira, and Kris,
>the Kraits

Kissable Keeper!

I am the national symbol of New Zealand
    and unique in the bird world.

Like a mammal,
    my bones are filled with marrow
    and my feathers look like fur.

I have an acute sense of smell
    because my nostrils are at the end
    of my very long beak.

I have the lowest body temperature
    of all my relatives!

So, you see, unique is good.
Good in me and good in you!

Kinder and kinship,
Kahalil and Keith, the Kiwis

Kind Kalanchoe,

No matter what you look like,
    you are so beautiful!

It doesn't matter if you are
    pale white or deep black,
    every color is gorgeous.

I am a pallet of color.
I have light blue-gray feathers,
    bright, red-orange legs,
    red eyes,
    dark and light striped wings,
    and the only bird in the world
    with nasal corns (coverings)!

So, be colorful.
No matter what your color is!

Keyed to knowing you,
Kathy and Kristen, the Kagus

L

Look who's laudable!

Just like you, I have a lovely body.
But my greatest gifts are my ability
    to mimic any sound, I hear.

A lot of birds can do this
    but I am exceptional.
I can copy any sound,
    including musical instruments
    and power tools!

My Wisdom to you is this.
Listen.
Learn.
Repeat what is helpful.
Discard what is not.

Lovingly,
Lois, the Lyrebird!

Learned Lady!

I am an aquatic snail called a Limpet.

My best feature is my teeth.
I have the strongest teeth in the world
    13 times stronger than steel.
I keep them strong by replacing
    each row of my hundred rows
    every 47 hours.

So, brush your teeth
    to keep them strong.
Stretch your muscles
    to keep them strong.
Expand your mind
    to keep it strong.
Then you can bite into anything you want!

Lucky to know you,
Levi, Liam, and Logan,
    the Limpets

Loveable Lass!

I am a Lemming
      one of the smallest rodents.

Like you, I am amazingly cute
      with long, soft fur.

No hibernation for me —
      I am active all year
      in my home at the Arctic Circle.

I love solitude.
In fact, I only pair up to mate
      which I can do only one month
      after my own birth!

Laughingly lavish,
Linda, Lisa, Liz, and Lysset,
      the Lemmings

Mindful Mate!

I am a member of the leopard family
    complete with huge paws,
    a long tail, and
    very cool stripes and streaks
    of color.

I can climb anything
    as I am the most adaptable cat
    at scaling
    in the world.

Do what I do:
    wear your distinction with honor
    and
    climb everything in front of you.

Meritoriously,
Mandy, Maria, Marion, Marsha,
Marianela, Mary Kay, Mary Sue,
Mary, Melanie, Michelle, and Millie,
    the Margays

Masterful Munchkin!

I am the largest of the deer family
    and the tallest animal
    in North America.

Despite my size, I can run 35 mph!
My impressive antlers can span six feet
    and I grow new ones each year.

My Wisdom is:
    know you always have choices.
Choices about how to think,
    how to act,
    how to respond.

Then, choose wisely.

Madly yours,
Matt, Max, Mazen, Michael,
Mike, and Moses,
    the Moose

Majestic Mystery!

I am not your ordinary goat.
Most goats appear kind of silly.
Not me.

I am so majestic
      that when I stand
      on a mountaintop,
      with my full beard
      and corkscrew long horns
      I exuded Wisdom.

So, stand tall.
Show off all your gifts
      and share all your Wisdom
      with pride.

Modestly,
Mary, Mary Anna, Mary Lou,
Mary Lu, and Melissa,
      the Markhors

Natural Novice!

I am part of the most numerous animals
   on the planet.
We account for 8 of every 10 animals!
I am the Nematode or roundworm.

We can be as small as 1/10 of an inch
   or as long as 28 feet.

Like almost every creature,
   we can be helpful or hurtful.
The world is full
   of colorful characters.
Some helpful, some hurtful.
Some docile, some imposing.

Use your insight and intuition to discern
   what is useful and what is not
   in this wonderful world.

Nostalgically,
Nancy, the Nematode

Noticed You!

I am a Natterjack,
    a toad
        who lives throughout Europe.

You probably know about me
        from fairy tales
            where I turn into a prince.

That is my message to you.
Be transformative.
Change what you need to change.

Life is too short to waste
        being someone you don't like!

Nutty about you,
Noah and Nick, the Natterjack

Newest Noble,

You are the newest being to the planet;
I am a living fossil of one of the oldest—
   present for 500 million years.
From the Greek word for sailor
   I am a Nautilus, a marine mollusk.

I bring the message that life is change.
Everything changes.
Always.
So, keep your hand open.
Refrain from clinging or
   hanging on too tightly to anything.

That way, when change comes,
   and it will,
   you will be ready.

Next to you always,
Nathan and Ned, the Nautilus

Outlandish One!

My best Wisdom is: be yourself.
Your body is fantastic, just the way it is.
Your mind is brilliant, just the way it is.

You see, I am strange looking.
I'm the only living relative of giraffes
      but I have stripes on my legs
      like zebras.
Weird, right?

You, too, will have to learn to
      accept yourself as you are.

Because, no matter what,
      you are the best you there is!

Optimistically,
Opal, the Okapi

OOOOOOOO

I am known for my wisdom.
Perhaps this is because
      I can turn my head 270 degrees!
This means I look at all angles of life.

I hope you do the same.
Look forwards and backwards,
      left and right,
      up and down.
      and all angles
And be sure to look inside as well as out!

Openly searching,
Otto, the Owl

Optimistic One!

As an Ocelot,
    also called the painted leopard,
    I am part of the cat family.

Like most cats, I am a solitary being.
That is my Wisdom to you.

Love yourself.
It is essential to cherish yourself
    before you can love another.

Enjoy being with yourself.
If you don't like yourself,
    no matter who else likes you,
    their love won't be enough.

Embrace solidarity every day.
At least for a few moments.

Overjoyed,
Olivia, the Ocelot

P

Pioneering Peach!

Everyone recognizes me immediately!
I am a peacock
    one of the few species in the world
    that is blue.

If I am a male,
    my tail feathers are six feet long
    and take up 60% of my body length.
That is so I can attract a partner.

My lesson for you is:
    be yourself to attract others.
You don't have to be fake.
Or even blue!

Permanently in with you on the path,
Patty, the Peacock

Peaceful Pacific!

I am a deep-sea diver by profession.
I live in a very cruel environment
    in my native Artic,
        where it is often -40 below zero.

I dive for my food,
    down to depths of 1,755 feet!

You also may have to work hard in life.
That's okay.
You will learn to dive deep,
    despite any harsh environments,
    to thrive
    despite any harsh conditions.

Passionately,
Paul, the Penguin

Phenomenal Person!

I am a Parrotfish.

I am very colorful
    with a dramatic row of sharp teeth
    that allow me to carve algae
    off coral reefs.

Because of this feature,
    I can live in severe conditions
      where food is difficult to get.

You are like me!
You can master any environment
    and overcome anything!

Trust your natural abilities
    to flourish and prosper.

Pleasantly,
Pat, the Parrotfish

Qualified Quartz!

I am a Quail, a game bird.
I share with you my favorite pursuits:
    family and adventure.

Family is vital.
And family can be blood or chosen.
Because "family" is anyone who
    cherishes and respects you.
"Family" is anyone who
    knows who you are
    and shares who they are.

My family and I travel together
    worldwide.

Be willing to go anywhere
    on this fascinating planet.
Explore it with curiosity and wonder.

Quilted in love,
Quinn, the Quail

Queen of my Heart!

I am a Quokka
    the smallest of the marsupials.
A fancy way of saying "pouched animal."
I reside in Australia and New Zealand.

I am very social and friendly.
And darn cute!

I like to make my own runways
    through the long grasses.
Even if other avenues open.

You can do the same.
Make your own path —
    wherever you want!

There is always another route
    even if you must create it!

Quenchless in my devotion to you,
Quynh, the Quokka

Quintessential Quince!

As a Quetzal, I am considered
	the most beautiful bird
	in the world.
Colored bright red, green, black,
	brown, white, or yellow
	with 40-inch blue tail feathers,
	who am I to argue?

If others say great things about you,
	let them.
Your job is to be humble.
Humility is knowing the truth
	about yourself.
The great things and
	the not-so-great things.

Humility is embracing your assets
	and owning your flaws.

Quaky about you,
Quincey, the Quetzal

R

Robust Red Rose!

I am an intelligent and clever animal
    with a black mask and ringed tail.
Some folks consider me a nuisance
    but I am fascinating.

I am the symbol of ingenuity.
There is no problem
    that I cannot address.

So it is with you.
For every issue, there is always
    a creative solution.
Use your ingenuity and insight.

Sometimes, the answer is
    to do something external.
Sometimes the answer is
    to do something internal.
Like shift your attitude!

Rambunctiously Yours,
Roberta and Robyn, the Raccoons

Radiant Redbud!

My name is Rhinoceros
      Greek for "nose-horn."
Be glad you don't have a schnoz like mine.

I am on the critically endangered list
      because of human greed.

Please, please, please!
Take care of all my people.
And my people are the four-leggeds
      and the winged ones
      and the swimmers
      and the crawlers
      and the rock people
      and the many layers of earth
      and of course, the two-leggeds.

Reverently,
Robert, Rodnick, Roger, and Ryan,
      the Rhinoceros

Ravishing Ranunculus!

I am a designer dog
      called a Raggle.
A rat terrier and beagle combo.

My family tree is very intelligent
      and energetic and alert.
So am I.

But I am also very stubborn
      and hard to train.
My ego gets in the way.
I want to do everything my way.

If you have this trait, learn from me.
I don't know everything
      and neither do you.
Be willing to tame your stubbornness
      and learn new tricks.

Rigorously rooting for you,
Reagan and Ruth, the Raggles

S

Sacred Soul!

I am a member of the lemur family.
That means I am related to monkeys.
We all have short snouts and long tails.

And I glide!
I can spread my arms
and sail from tree to tree.
It's so cool!

My Wisdom to you:
      spread your wings and let go!

Squarely in your corner,
Shawn, Shelly, and Stacie,
    the Sunda Colugos

Sweets!

Never be held by the norms of society.
You can break them just as I have done.
Be bold!

I am unique. As you are.
I am a male but I get pregnant
    and give birth to my young.
My pregnancy only lasts 20 days
but I deliver 2,000 of my offspring.

You may not break convention like that
    but you will, at some point,
        have to stand against the crowd.
This takes bravery and confidence.
Don't worry —
    you have both!

Surprisingly Sentimental,
Steve and Scott, the Seahorses

Superior Sunshine!

I bring you this message: relax!
Go slow. Enjoy the day.

I do this so well
It's even in my name: Sloth.

I am notoriously slow.
I don't move much
    and when I do,
      I take my time.
In fact, I only poop once a week.
I bet your parents would like that!

So, slow down.
Cherish the moment!

Smiling,
Sandy, Sam and Sue, the Sloths

T

Tremendous Talent!

I am a turkey.
And not just your average turkey.
I am a Wild Turkey.

I am often viewed as the underdog
    (or in this case, underturkey)
But I am a very spiritual being.
I bring you the gifts of
    acceptance and surrender.

Don't worry if you're the underdog.
Trust.
Do the next right thing.
Do the best you can at this moment.
 Then let go of the outcome.
    and accept that life is unfolding
    as it should.

Thankfully,
Tara, the Wild Turkey

Top of the morning to you, Terrific Tot.

I am one of the smallest animals
    at only .02 inches in length
      but I can survive anything!

I can live at the bottom of the ocean
      and on top of Mount Everest.
I have even been to outer space.

You, too, can make it anywhere.

You can be
    anything,
    anyplace,
    anytime.

Tastefully Transcendent,
Terry and Tony, the Tardigrades

Tough and Tender Child

I am a Takin
>   a relative of sheep and goats.

I have unique features allowing me
>   to love the harsh environment
>   of the Himalayas.

First, I have a complex sinus system
>   that warms the air
>   before it gets to my lungs.
Second, I wear two coats in the winter!

Please, put on extra layers!
Yes, this might mean extra coats.
But it might mean extra courage,
>   or extra patience,
>   or extra understanding.

Tenaciously thoughtful,
Todd, the Takin

U

Unbridled Uniqueness!

I am a Uguisi
    a beautiful small bird of Asia.

I am treasured for two characteristics.
First, I have a beautiful voice
    and I love to sing to the world.

Second, believe it or not,
    my poop is used in facial creams!
Don't exactly know
    how that was discovered
    but there it is!!
All the big cosmetic firms want me!
Wouldn't it be great
    if your poop was worth millions?

United in Love,
Una, the Uguisi

Uncommon Unconventionality!

I am a monkey of South America.

I stand out because I have red face.
Not just when I get embarrassed.
No, I have a permanently red face.

So, when you get embarrassed
    and your face turns red,
      think of me.
It's okay to be embarrassed.
Own it. Admit it.
When possible, laugh at it.

Unlike me, your face will return
    to its normal color
    and life will go on.
I promise!

Ultra-devoted,
Ulysses, the Uakari

Ultimate Soul!

I am an ancestor of the modern sheep.
Not domesticated, I am wild!

Although I can see just fine
    I communicate by scent.
My message to you is
    remember to use all your senses.

Each sense broadens and enhances
    your knowledge of this world.
Ignore a sense and
    you may miss something vital.

So, look, listen, taste, smell, touch.
Explore and investigate everything!

Unrestrained devotion,
Uriah, the Urial

V

Vivacious Vision!

I am a Vicuña
    an animal of the Alpaca family.
As the national animal of Peru,
    I live in the Andes Mountains
    throughout South America

My wool is the most expensive
    because it is the softest
    of any animal in the world.

My Wisdom: be soft.
Be gentle, kind, tender.

True power comes from knowing yourself
    not from running over others.

Victoriously Yours,
Vienna, the Vicuña

Venerable Vessel!

I am a veery
    a small migrating bird
    of the Western Hemisphere.

As I navigate from the arctic Circle
    to the southern tip of Brazil,
    I fly 160 miles a night!

For your long adventures
    take with you the gifts I employ:
        strength
        dedication
        awareness
        endurance.
And never lose sight of the goal!

Voraciously cheering for you,
Victor, the Veery

Valorous Valiant!

I am not the most popular critter.
In fact, I look scary and unattractive.
As an arachnid, I am kin to spiders.
My official name is Vinegarroon
       but folks call me a whip scorpion.

After mating, I retreat
       to a hidden location for months.
Then I lay 30 to 40 eggs
       which I brood for two months.
When my children hatch,
       they ride on my back
       until they're independent.
I do not eat this entire time.

Be willing
       to sacrifice yourself for others.
Help another become interdependent.

Validating you always!
Vanessa, the Vinegarroon

W

Welcome to this Wacky World

I am a Wombat!
A fuzzy, calm, cuddly-looking marsupial
 or pouched mammal.

I have two unique features.
One is that my pouch faces backwards.
Two, my poop is cube shaped!
Can you believe that?

Your poop may not be shaped like mine
 but you too are fascinating.
Just like me!!

Don't be embarrassed
 by your uniqueness
 even if it's your poop.
It's what makes you, you!

Warmest Wishes,
Wanda, the Wombat

Way to go, Wonderful Wild Windflower!

As a Walrus, I live all over the world.
Everywhere!
And I am content on land or in water.
I bet you are too!

My huge tusks are spectacular.
I use them to icepick my way out of ice,
     to help haul my big body onto land,
     and
        I stab them into ice
        so that I can float in the water
        when I want to take a nap.

Pretty cool.
My advice: use what you've got!

Wealthy in Warmth,
Wade and Willie, the Walruses

Wise Wisteria,

You got here
    to this beautiful and harsh world.

As a Wasp, I symbolize the harshness:
    biting and stinging easily and often.

You will need
    courage to face me
    strength to heal from my stings
    determination to move forward
        despite the possibility
        of getting stung again.

Your reward will be untold growth
    to be the master of your life.

Wildly welcoming,
Warren, the Wasp

Some fun facts for the letter "X."

There are:
817 million species of animals on earth[2]

30 million species of insects
And some 10 quintillion
(10,000,000,000,000,000,000)
individual insects alive[3]

4,00 species of fish[4]

8,100, 903,329 humans
at the time I looked this up[5]
and of course,
you make one more!!

---

[2] https://www.worldatlas.com/articles/how-many-animals-are-there-in-the-world.html Accessed Dec. 5, 2023.
[3] https://www.si.edu/spotlight/buginfo/bugnos Accessed December 5, 2023.
[4] https://a-z-animals.com/blog/how-many-fishes-are-in-the-world/ Accessed December 5, 2023.
[5] https://www.worldometers.info/world-population/ Accessed April 1, 2024.

y

Yikes!

My name is "Yabby"
    which means "Crayfish."
I live in Australia
    where I can survive long droughts
    by burrowing into the soil.

I am the symbol of
    determination and abundance.
Avoid a major fear:
    that this is a planet of scarcity.
Realize a major truth:
    that this is a world of abundance.

Shortages and plenty ebb and flow.
Learn to ride the waves
    of giving and taking
    of acquiring and releasing.

Yearning for your success!
Yuleima, the Yabby

Yowzah! Yawzah! Yawzah!

I am a viper snake of South America.

As a Yarara,
    I shed my skin
    as body grows too big
    for my old skin.

My Wisdom to you is:
    do the same.

Don't be afraid to let go of old "skin"
    when you have outgrown it
    or
    it no longer serves its purpose.

Yielding to greater Wisdom,
Yvette, the Yarara

Young Yellow Yarrow!

I am a huge, wild looking mammal
    with crazy fur and imposing antlers.
My name is Yak.

I love it at 20,000 feet
    in the high mountains of Tibet
    and enjoy temps down to -40.

I have much ancient Wisdom
    and
    share that Wisdom eagerly!

I am here to remind you
    that we all have a higher purpose.

Seek yours!

Yours always and forever,
Yoshi, the Yak

Z

Zesty Zephyr!

I am a Zonkey!
A hybrid.
My parents are a zebra and a donkey.
Pretty cool, right?

My back and torso look donkey-like
    with solid tan or brown coloring.
But I have striped legs like a zebra.

I have incredible strength and
    can carry 125 pounds on my back.

My Wisdom to you:
    carry your load
        with power and grace.

Your zany friend,
Zelda, the Zonkey

You're in the Zone!!

I am a member of the rodent family.

My friends and I are masterful diggers
    and carve tunnels
      hundreds of feet long!

Although no one sees our labor
    we are very industrious.

That is my message for you.
No matter if you work in the spotlight
    or behind the scenes,
      do your best.
Take pride in your labor.
Work is a right
    to be savored and enjoyed!

Zigzagging all around you,
Zack, the Zokor

Zen Zantedeschia!

I am a breed of cattle called a Zebu
    whose ancestors first appeared
    over 8,000 years ago in Asia.

I am considered sacred in Hinduism
    and have been pictured on coins,
    statues, and architecture.

You too, are sacred —
    as is everything in creation.

Learn from all creation.
Everything is part of the Whole.
Everything has Wisdom.

Learn the Wild Wisdom around you.
Share your Wild Wisdom with others.

Zealously yours always,
Zanzibar, the Zebu

www.ingramcontent.com/pod-product-compliance
Lightning Source LLC
Chambersburg PA
CBHW050440010526
44118CB00013B/1615